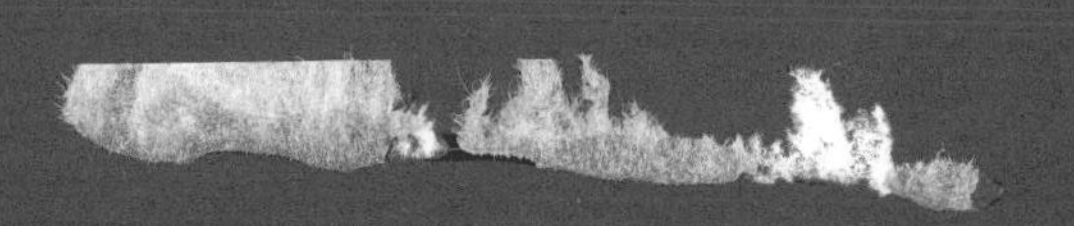

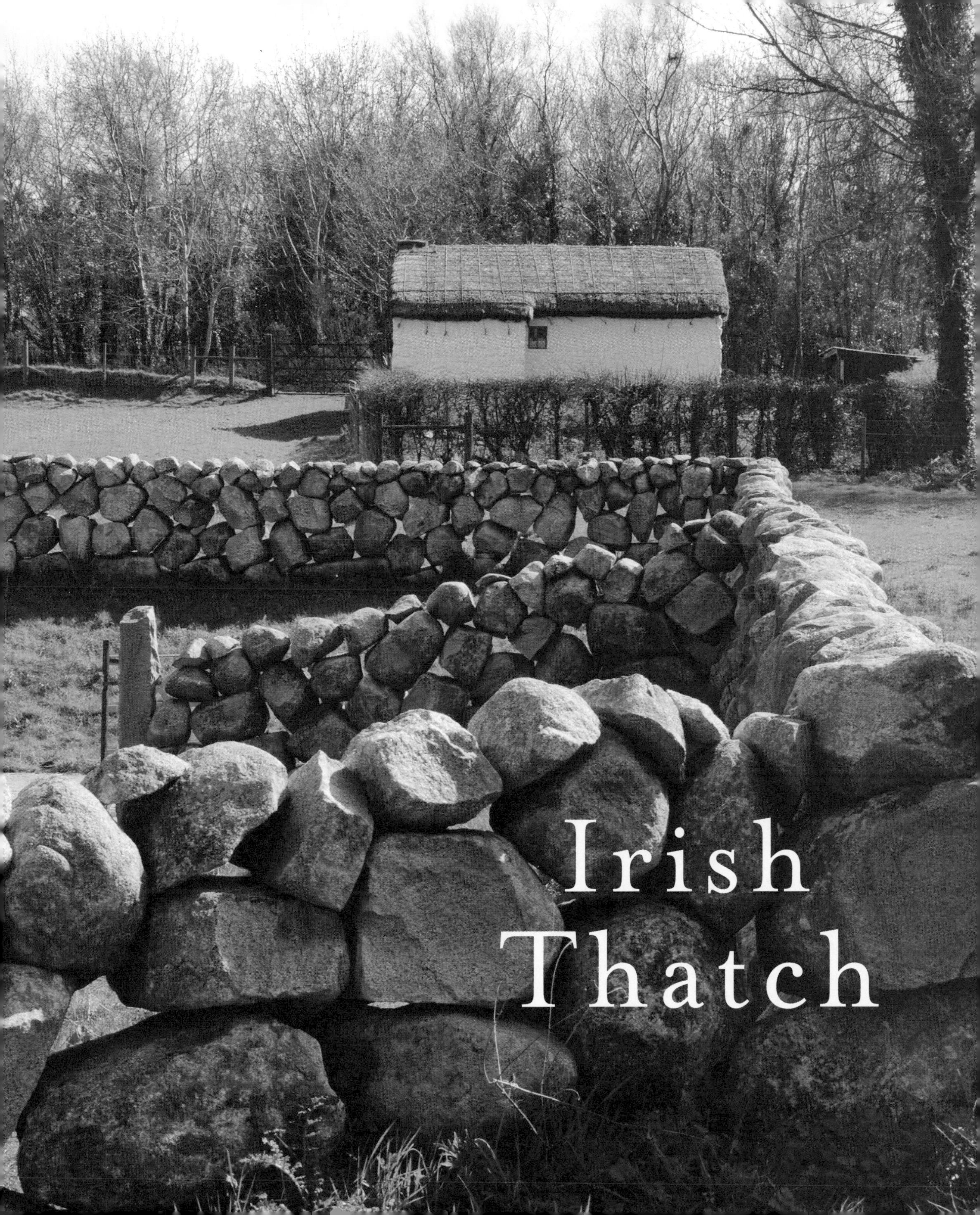
Irish
Thatch

Patrick Pearse's Cottage, Connemara, County Galway.

DEDICATION

This book is dedicated to the many who take care of these houses for posterity.

Irish Thatch

EMMA BYRNE

THE O'BRIEN PRESS
DUBLIN

Pretty thatch, Dunmore East,
County Waterford.

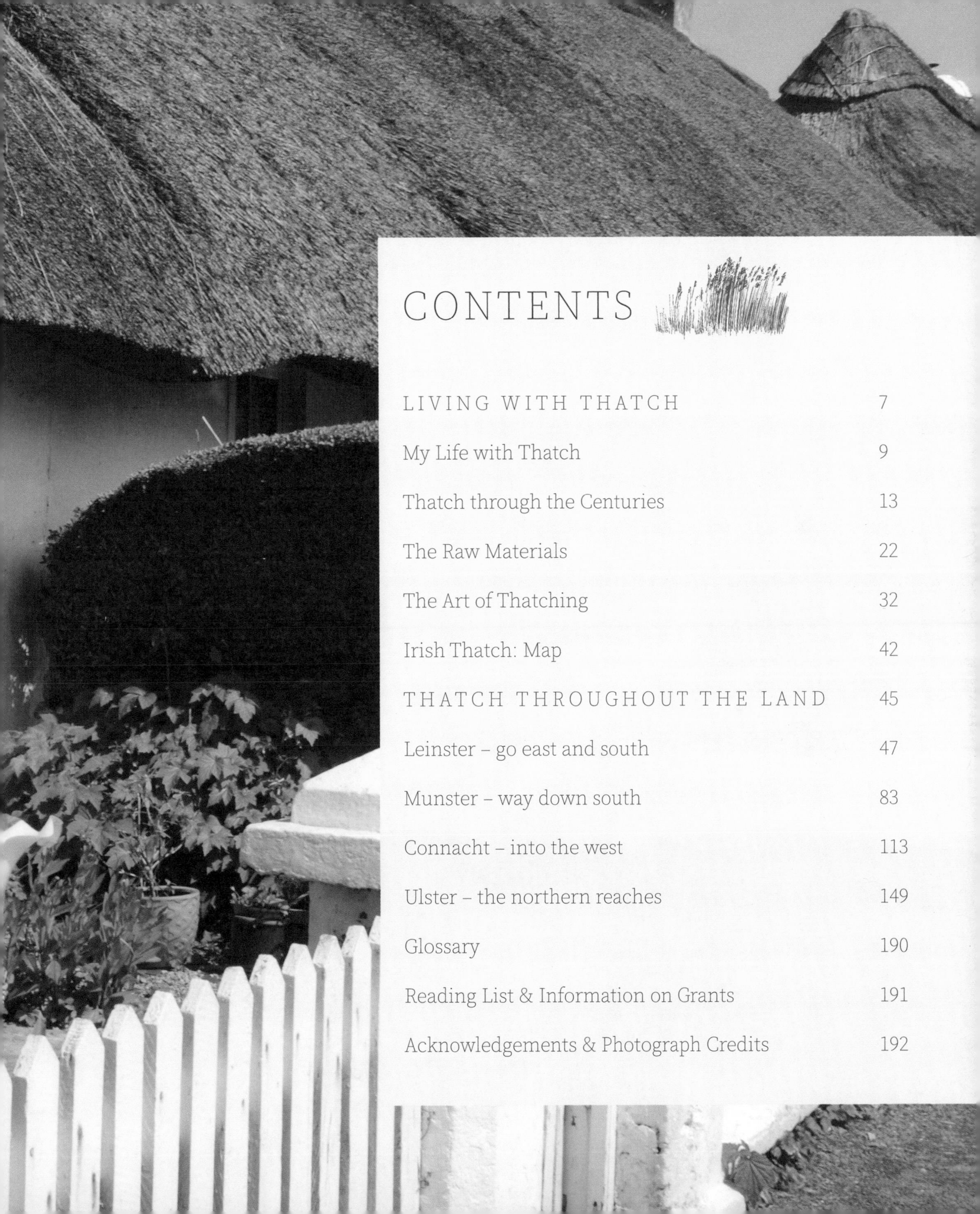

CONTENTS

Killannin, near Oughterard, County Galway.

LIVING WITH THATCH

Above: Old Mill Cottage, home to the author and her husband Jonathan **(below)**.

MY LIFE WITH THATCH

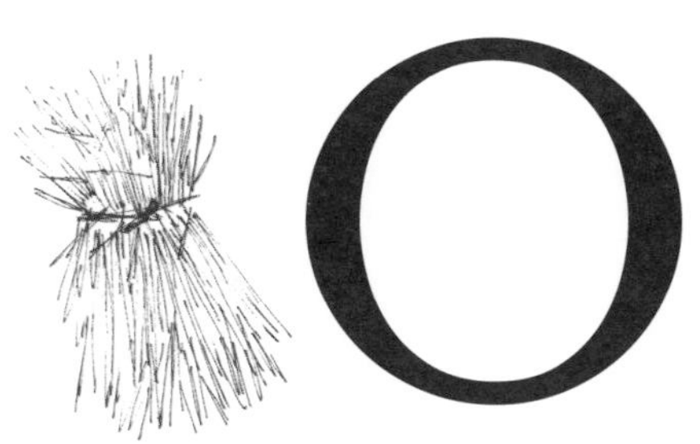

One of the most iconic images of Ireland is the thatched cottage. Whether in romanticised films such as *The Quiet Man* or on all kinds of souvenirs, it is forever tied up with the Irish national identity.

Many people may believe that these dwellings are few and far between, but when I began travelling the country I was surprised at just how many are left.

So why did I become interested in thatch? That's simple: I bought one. I was house-hunting in County Wexford when I came across a beautiful thatched cottage on the edge of a small village and thought, this is the place for me (and my husband, Jonathan, when I convinced him). Nothing matches the feeling of warmth and comfort that comes from three-foot-thick clay walls and a two-foot-thick natural roof, keeping in the heat of your blazing stove, or knowing that your two-hundred-year-old house has stood for so many years.

All houses have their own character and needs, and a thatched cottage has its own particular requirements and style. Birds try to eat your roof; rodents burrow into it to keep warm; clay walls are damper than modern ones; interiors may seem intimate to some, but pokey to others. In addition, there are unusual financial costs: house insurance is both difficult to obtain and costly, and re-thatching can

Above: A field of wheat, just ready for harvest. **Below:** Close-up of a reed-thatched roof.

be very expensive. (Grants are available, but the process is long and complex.) Many are also listed buildings, so the exteriors cannot be altered without official permission. Previous owners may have made alterations that need attention; plumbing and electrical wiring may need overhauling. One especially memorable moment occurred when I was woken up by water dripping onto my face from the bedroom lightbulb! Living in a thatched cottage, one soon learns to cope with the unexpected.

However, thatched houses look to the future as well as to the past. The fact that they are constructed using all-natural materials means that they are environmentally friendly. The houses are built from clay and stone, while the thatch is fashioned from sustainable materials such as water reed, which grows in areas that are otherwise agriculturally unviable, or straw, which is a by-product of food production. By their design, these dwellings are extremely well insulated, easy to heat, and respond well to the changing seasons; they are warm in winter and cool in summer, unlike many modern houses. Their appearance also blends well with the local landscape, and is one of the main reasons these buildings are so photogenic. Maintaining a thatched house also helps to keep alive the wonderful craft of thatching and provides a living for the skilled men (and a few women) who practise it.

In the first section of this book, I will briefly explore the history of

thatch in Ireland, examine the raw materials involved, and explain how a skilled thatcher carries out his work.

The second section takes the reader on a tour of thatched buildings throughout the country, from bustling city streets to remote boreens with grass growing in the middle.

From Glencolmcille in Donegal to Kilmore Quay in Wexford, armed only with a camera, a notebook, and an invaluable Sat Nav, I have endeavoured to include a cross-section of different thatching styles, from ancient to modern, as well as the quirky and charming little houses that caught my eye as I drove past. Unfortunately I have not been able to fit all the thatch of the land into the book, and the culling process was a difficult one.

Irish Thatch is by no means intended to be a definitive or complete guide to thatch, nor an architectural or conservation study, of which there are many excellent examples elsewhere. Instead, I see this book as a celebration of a form of building that has survived from ancient times to the present day – a part of our shared history and part of our future. My hope is that others may feel inspired to explore and learn to love this unique and important aspect of our heritage.

The thatcher working on the gable of Old Mill Cottage.

The metalworker's house at the Irish National Heritage Park, County Wexford.

THATCH THROUGH THE CENTURIES

Thatch has been used as a roofing material in this country for over nine thousand years, making it one of the oldest building crafts employed in Ireland. Prior to the introduction of other roofing materials, such as tiles and slate, in the eighteenth and nineteenth centuries there was little else, except turf and sods, which were sometimes combined with thatch.

Thatched roofs were mostly used on what is known as 'vernacular' (or 'word-of-mouth') buildings. A vernacular building is one where ordinary people build their own houses, barns and stables from local resources, independent of trained architects, formal styles and fashionable trends.

Amongst the current main thatching materials are wheaten straw, oaten straw, rye straw, flax and water reed.

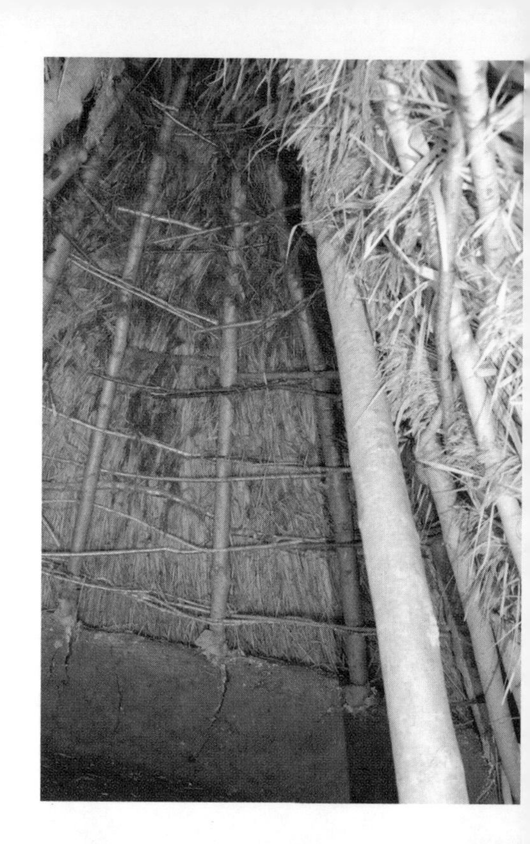

MESOLITHIC OR MIDDLE STONE AGE

Very little is known of our built heritage from the Mesolithic or Middle Stone Age period, 9000BC. However, in the Irish National Heritage Park, Ferrycarrig, County Wexford, there are reconstructions of ancient remains found at Mount Sandel in County Derry, and other places.

These dwellings could have been thatched with river reed and hazel stays (strips of wood) that helped to hold the thatch in place. The frames of the reconstructed structures are of local wood, filled out with mud walls. Reed was the main material used up until the tenth or twelfth century. It is also possible that some of these dwellings may have been tepee-like in shape.

Top right: Notice the mud base on the walls and the rough-hewn condition of the timber branches. **Below:** These basic homes could have had distinctive shapes. You can see the leather drying outside; the 'fireplace' is also outside.

NEOLITHIC OR NEW STONE AGE

Neolithic or New Stone Age homes from around 6500BC went up a notch as people became more settled farmers. Still thatched, it is thought they were more rectangular in shape (**see below**), made of oak planks and wattle. The roofs were high pitched and had two or three rooms. These dwellings were more comfortable, with separate places to sleep and eat. You begin to get a sense of the habitable space that was the early precursor to our homes as we know them today.

Right: The fireplace is now within the home. There is a sleeping area to the left. **Left:** The window allowed light in and smoke out. **Below:** These buildings were quite tall and some archeologists believe that they were feasting houses rather than homes.

THE BRONZE AGE

From about 2300BC until AD600, the Bronze Age brought greater learning and therefore more prosperity to Ireland. The 'Beaker' people arrived and with them the new wonder of metal. First there was copper, then bronze and eventually gold. This period saw the emergence of defensive ringforts, to control wider territories. These buildings were also probably thatched.

Ringforts were circular areas enclosed by an earthen bank and a ditch. The walls were built with stone or wood. They were thought to be the farmsteads of the wealthy, with poorer people living outside in open settlements.

Brehon laws tell us that inside the structures there was a family house, which had stone walls, an outhouse, animal pens, often beehives, and farm equipment such as a plough, spades, billhook and threshing sticks for corn. Many ringforts also had a souterrain (an underground passage or chamber) for storing food or as a refuge from attack.

Opposite top left: The ringfort had an impressive entry point. **Opposite top right:** Wooden and wattled walls are replaced with stone. People have metal pots for cooking. **Opposite bottom:** Only the animal byre is now made with wattle and timber. **This page, top right:** Notice the *súgán* (straw ropes) used to stabilise the structure. **Below:** The interiors look much more comfortable and warm. The Irish National Heritage Park allows vistors to stay and experience life in a 'medieval ringfort'. Visitors are given costumes and food of the period.

CHRISTIANITY

With the coming of Christianity, other roofing materials were used, particularly for sacred buildings like the church and scriptorium. The more secular buildings where the monks ate and slept may well have been thatched. Some of the larger monasteries developed into streets and towns, where the buildings would have had thatched roofs.

All of these buildings are reconstructions at the Irish National Heritage Park, County Wexford.

THE VIKINGS

The Vikings began arriving in Ireland from around AD800. These settlers, originally from Scandinavia, retained the art of thatching for their buildings. Viking ships made it to America long before Christopher Columbus, and they travelled east to Byzantium. They brought a significant culture to Ireland, introducing the first coins and new art. They established many of Ireland's towns such as Dublin, Wexford and Cork.

THE CRANNÓG

This was also the period of the crannóg, an unusual form of Irish settlement. An artificial island was created in a lake, made of layers of stone, clay, timber and even bone. Some crannógs had been around since the Bronze Age and some continued to be used up until four hundred years ago, when they were Gaelic stongholds against the British crown. The main buildings were probably circular with a cone-type roof, which was thatched.

In these settlements one can see the ancestor of the clachan (from the Irish *clochán*), or small village grouping, some of which survive today, and are pictured in this book.

DIRECT-ENTRY AND LOBBY-ENTRY HOUSES

From late medieval times through the seventeenth, eighteenth and nineteenth centuries right up until today, thatched houses fall into two broad categories:

Direct-entry houses, where the door opened directly into the main living space. Lobby-entry houses, where the door opened into a small lobby, giving way on either side to the main living space and a bedroom.

DIRECT-ENTRY HOUSES

These houses were mostly seen in the west and north. They typically featured rubble stone walls and rounded thatched roofs secured on stone or timber pegs, which stood up better to Atlantic winds.

It was common to have two doors opening directly into the kitchen on opposite sides of the house. Which one to use was based on which way the wind was blowing. The front door incorporated a half-door, useful for keeping animals out and children in. In these houses there was a nook or bed alcove very near the hearth, big enough to take a double bed for the most senior members of the house. A curtain or screen was placed around this for privacy.

Near Glencolmcille, County Donegal.

Kilmuckridge, County Wexford.

LOBBY-ENTRY HOUSES

In these houses the door was positioned in line with the hearth with a screen wall built in front of the door to protect the inside from view. As with direct-entry houses, the hearth was where all the food was prepared and where people gathered in the evenings to talk and tell stories. These lobby-entry houses were mostly found in the midlands, east and south-east of the country. The thatch was made of locally sourced cereals and was often hipped or half-hipped (see glossary), whilst in the north and west the roof was more often gabled.

THE ONE-ROOMED WORKMAN'S COTTAGE & IRISH BYRE HOUSES

The one-room workman's cottage, often used by migrant workers, was the most basic thatched dwelling and was sometimes shared by animals, a common practice because animals, cattle in particular, were a valuable asset. These Irish byre houses were often built on a slope so that the discharge from the cows would fall away. The highest part of the sloped building was where the people lived.

Mountcharles, County Donegal.

Below: A new housing estate in Kilmore Quay, County Wexford, is very much in keeping with the thatched aesthetic of the village. Most of the surviving Irish thatched properties date from the seventeenth to the nineteenth centuries. However, there are also new builds, such as this housing estate, and holiday homes. Thatch is less demanding on the land as the raw materials do not require quarrying or mining. The materials are sustainable and biodegradeable.

THE LANGUAGE OF THATCH

The art of thatching has a language all its own. For explanations of terms, see the glossary.

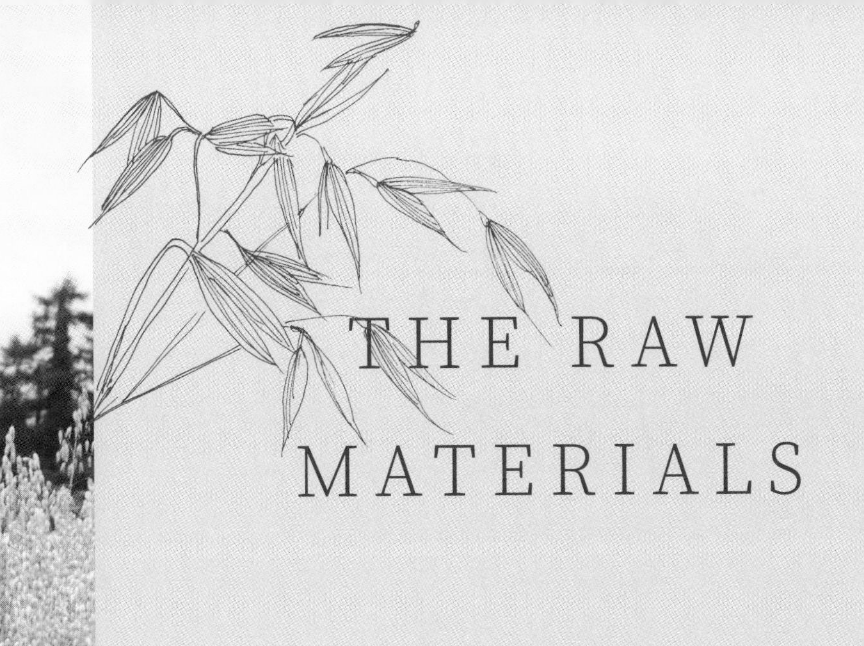

THE RAW MATERIALS

Contemporary thatching materials include wheat, oats, barley, rye and water reed. Most buildings in Ireland use wheaten straw or reed. Thatch made from reed has the best longevity, although most reed is now imported due to concerns over nitrogen levels in Irish river water.

Other materials have been used, often based on what is available locally, for example: marram grass in coastal areas; flax in Donegal and Northern Ireland, and rye straw in the west of Ireland and on the Aran Islands.

However, there have been roofs made using creeping bent grass, heather, saw sedge, broom, yellow flag iris, water reed, bracken, black bog rush, bulrush, gorse and even potato stalks.

The other notable natural materials used in thatching are hazelwood and willow. These wooden rods are used as stays to secure the thatch in place.

WHEAT STRAW

Technological change in farming has had a significant impact on wheaten straw. The availability of good-quality thatching straw declined after the introduction of the combine harvester and the release of short-stemmed wheat varieties. The increasing use of nitrogen fertiliser from the 1960s also weakened straw and reduced its longevity. Since the 1980s, however, there were some moves back to the older, tall-stemmed 'heritage' varieties of wheat such as Squareheads Master and N59.

My own house is thatched with N59, grown in County Meath by a thatcher who also supplies other thatchers throughout Ireland. It is cut in a traditional manner (with a converted rice cutter), and threshed, to avoid being broken by a combine harvester. Other thatchers use a reaper and binder. All straw is wetted and pulled before thatching to ensure its longevity.

Main: A field of ripe wheat, a few weeks before harvest. **Inset:** When the ear of wheat bends like this, it is ready for harvesting.

OAT STRAW

Oat straw is not as long lasting as wheat, but it is still used as it's easier to source than wheaten straw. Before the regrowth of heritage varieties of wheat it was in common usage. It is a very pliable material and easy to work, making it a favourite with thatchers. Common varieties of oat that produce good thatching straw include potato oat and black oat. Oat straw has a distinctive yellow colour.

RYE STRAW

Rye straw was grown in the west of Ireland and the offshore islands, as it had a good tolerance for salt. Rye straw is as durable as wheat and is also long, soft and more flexible than any other straw. In the past, rye was grown on some farms as a specialised crop for thatch and was often grown in rotation with potatoes. It weathers well into a lovely slate grey or silver colour and can still be seen on some properties on the Aran Islands. However, there is little rye straw grown in Ireland today, so if it is used it has to be imported.

Above: A close-up of oat florets. **Below:** Rye roped thatch, on the Aran Islands, has a wonderful silver appearance when weathered.

HARVESTING THE STRAW

Traditionally, harvesting was an occasion that involved a large part of the community, whether it was wheat, barley or oats (**see left**). These days the work is done with a combine harvester in a matter of hours, but the straw harvested in this way is not suitable for thatching as the straw becomes too broken up. Many thatchers use older or alternative methods such as rice cutters, or machines that will not damage the straw.

In days gone by the corn was cut with a reaper and binder and bundled into 'stooks'. These 'stooks' were left for about a week to dry, then gathered into 'stacks', and after another week or fortnight the stacks were brought into either a shed, open yard or haggard, where they in turn were gathered into a large hayrick. This was sometimes thatched to protect the straw.

Towards the end of September or beginning of October the hay was then threshed in a threshing machine, which separated the straw from the grain.

Many rural harvest festivals still partake in a 'threshing', and it's usually a day for merriment for all members of the family.

Left: Different varieties of corn. Heritage varities of wheat, such as N59, have a much longer stem and less grain than contemporary wheat.

Opposite: A thresher, driven by a separate tractor, separates the wheat from the straw.

7325

GOLDCROP
IRISH
CERTIFIED
SEED
OATS
50kg
SINGLE DRESSED
GOLDCROP SEED
RTIFIED
ED BARLEY
50 kg

Reed by the River Slaney. Water reed thatch is popular wherever there are rivers, although nowadays it is mostly imported. Homes near the Shannon, Suir, Nore, Blackwater and Slaney are most likely to be roofed with reed.

WATER REED

Reed is probably the most durable thatching material available. The reed grows on marshland, producing golden stems five to seven feet high. A roof thatched with water reed lasts longer as the reed resists compression even when it is packed together tightly on a roof.

Some thatchers feel that excessive nitrogen levels in our rivers have adversely effected the durability of the reed. Recent improvements in water quality should reverse this trend.

Much of the reed is imported into Ireland, and due to its longer lifespan, reed has grown in popularity in areas that historically used straw for thatch.

Reed for thatching is cut into roughly three-foot lengths.

MUD WALLS

Many thatched houses have mud walls — especially in the south and east. They are made of wet earth that has been mixed with other materials such as straw, lime or gravel, to give the structure strength and make it durable. From ancient times right through to medieval times, clay and wattle (a mud wall with support) has been used throughout Europe.

Although it is slow to construct, as each layer has to be given time to dry, the end result can be as durable as stone. Mud walls make a building warm in winter and cool in summer. Buildings constructed of mud walls often blend into the landscape. In some places turf sods were also used as a wall material.

Usually the bottom of the wall has a plinth of stone, which keeps the wall dry. This is then covered with layers of mixed mud, one to two feet thick. The bottom of the wall can be up to three feet wide; it tapers out as it reaches the eaves.

Once the wall is complete, limewash is painted over it for protection and appearance.

Mud walls are still used in some countries in contemporary architecture as they offer high insulation, whilst also using local and sustainable materials.

Top right: Exposed mud walls, even though this property was in poor condition, the walls held.
Right: It is possible to make out other materials in the mud such as straw and gravel.

FIXINGS

For most thatched houses, a slender rod of hazel or willow, called a scollop (from the Irish word *scolb*), is used to fix the thatch to the roof. The thatcher uses straight and bent scollops. The hairpin shape of the bent scollop (*lúbáin*) acts like a staple.

Bent scollops are made by steeping the rods in water to make them supple, then 2- to 3-foot lengths are twisted into shape by hand. Freshly cut scollops need little or no soaking; dried scollops require overnight soaking. One leg of the hairpin is about 2 inches longer than the other, and is driven deep into the thatch, the thatchers using hand-leathers and wooden mallets. As many as 8,000 scollops may be used in the roof of a typical four-bay house.

In the past, scollops were prepared by the house owner. Some thatchers cut and prepare their own scollops; there are commercial producers in Ireland, but hazel scollops are also imported.

In the west and north of the country the thatch is tied down by ropes, hence the term rope thatch. Traditionally roofs were held down with *súgán,* or rope made from straw, but now other types of rope are used. The ropes are fixed to stone or metal *bacáns* that run along the gable end or the bottom of the eaves.

The scollops are cut into lengths of up to 6 feet and further cut into lengths of about 2 or 3 feet. These ones are made from hazel.

Examples of scollops:
Below: Ballyedmond, Gorey, County Wexford.
Right: Near Lough Corrib, County Galway. Hairpin scollops are pushed into the roof by hand, the thatchers using hand-leathers and wooden mallets.

Willow, grown near the rivers, is very flexible so is great in tight turns such as on gables, porch features and dormer windows. Briars, ash and laurel are also used.

THE ART OF THATCHING

TYPES OF THATCH

There are four types of thatching technique in use today in Ireland – each with its own regional variations. The most common type is **scolloped thatching**; the others are: **thrust thatching**, **roped** (with **pegged** as a variant of roped) and **stapple thatching.** All of these methods are used for straw thatch, but only scolloped thatching is used for reed.

Scolloped thatch: This is thatch that is held in place with rods or scollops made from either willow or hazel. Before the straw goes on the roof it has to be wetted and pulled to make it straight and to compact it into neat bundles. The hairpin scollops are driven in by hand or with a thatcher's mallet. Ideally, the scollops are horizontal or even upturned so that rainwater will not flow down into the roof. Once covered, the roof is beaten down and the eaves are trimmed with a shears.

Thrust thatch is mostly used in the north of the country and sometimes in south and east Leinster. Other terms for it include 'sliced', 'spliced' or 'fletched' thatch. It is mostly used with oaten straw and can last well depending on materials and workmanship.

The oaten straw is wetted and pulled in a similar fashion to wheaten straw. It is then bundled and the thatcher makes a knot in the bundle and thrusts this knot first into the roof. These bundles, often referred to as 'wangles', 'eaves', 'wads' or 'dolls', are tied a third of the way down their length and pinned to the eaves with wooden pegs. The thatcher continues working in layers, damping down and beating the thatch flat to form an even surface. Scollops are applied to the ridge and eaves, and the eaves are trimmed and the ridge finished.

Roped thatch is mainly found in the north and west as it works well against high winds. Traditionally, locally available thatching materials such as rushes and

Above: Roped thatch, Fr McDyer Folk Park, Glencolmcille, County Donegal.

Below left: Scollop thatch, County Tipperary. **Below right:** Exposed scraw or sods, the base layer that thatch goes over.

marram grass were used. The thatch is generally supported by a layer of scraw or sods, providing a good base for the thatch and an extra thermal layer.

The lifespan of a roped thatch is shorter than the wheaten straw equivalent by about four years, and the ropes need constant replacement because of the harsh conditions of Atlantic weather.

Pegged thatch is a slightly different version of roped thatch, and is largely historical as not many types survive. It was used in the northern counties and in parts of Connemara. In this method, timber pegs are pushed at close intervals through the ply of the network of straw ropes, further anchoring them to the thatch. In counties Antrim and Derry the ropes and pegs were traditionally of twisted bog-deal or pine from the bog.

Stapple thatch is associated with County Down and north County Dublin. It may be a variant of thrust thatching because tightly knotted bundles, or stapples, were used, but instead of thrusting the stapples into the underlying straw layer, they were held in place by lenses, or fillets, of clay (up to 3 inches thick) at the eaves, ridge and midway up the roof slope. A layer of clay was applied under the heads of the eaves course to establish the best pitch for the roof. At the top of the roof, bundles of straw were bent over the ridge and scolloped in. A capping of clay completed the ridge. The clay used for sealing the layers of thatch was fine, gritty mud found locally and mixed with cow dung to make it bond more firmly to the thatch. The thrust thatch technique applied to stapple thatch tends to result, over time, in a very thick and heavy roof, with each new addition of thatch and mud.

Traditionally, a simple bobbin ridge is applied. Various other types of copings, such as clay or timber boards, can be applied to the ridge.

Above and inset: Clay on the ridge of the house, near Salterstown, County Louth.

RE-THATCHING OLD MILL COTTAGE

Our house was built around the 1850s on the site of an earlier house that we were told dated from the period of the Wexford 1798 rebellion.

The photographs show how the house looked before it was re-thatched. The area around the chimney stack had already been patched and the ridge was very worn, as was the entire thatch, especially at the front.

Finding a thatcher proved a challenge. I searched the internet and heritage sites, and started to gather quotes. These differed by as much as seven thousand euro. Quotes in hand, we searched out the various thatchers' previous work as references. In the end, we found Peter Childs.

The complex process of applying for grants began. We ended up with financial assistance from Wexford County Council and the Department of the Environment.

Top and middle right: It is possible to see how the thatch has worn and shrunk. There are different opinions about wire mesh. It is said to prevent birds from building in the roof, but it rusts and the thatch shrinks from it over time. Several families of sparrows have succeeded in their architectural endeavours despite the wire. **Bottom right:** Old Mill Cottage just before re-thatching.

The scaffolding goes up and the stripping of the old and rotten thatch begins. The bundles of straw arrive and get set up and ready to go. The straw has been 'pulled' in the barn before it is bundled; this is an essential part of the process to ensure a strong thatch.

Above: The straw goes on in layers and is held in place with twisted hazel stays or scollops. Once the front and back are completed, the thatchers start on the sides, building up the straw as per the front. **Left:** A stook or bundle of straw. **Below:** The thatcher is building up extra depth to make a curved feature for the window, using the hazel stays to hold the thatch in place.

Right: When the front, back and sides are done, bobbins seal the top. They are made from twisted lengths of pulled straw and used to reinforce the thatch. To make a bobbin, a handful of clean, long straw is twisted at its centre and folded in two, and a loop is created. About fifteen bobbins are strung on a single stretcher. This is secured to the ridge with scollop pins.

Below: Once finished, copper sulphate, or bluestone, is applied. It stops moss and grass growth and protects against rot and rodents. It is recommended to spray annually.

Essential thatching tools include gloves, kneepads, mallets to drive home the 'stays', and thatching needles **(below and left)** for measuring the depth of thatch.

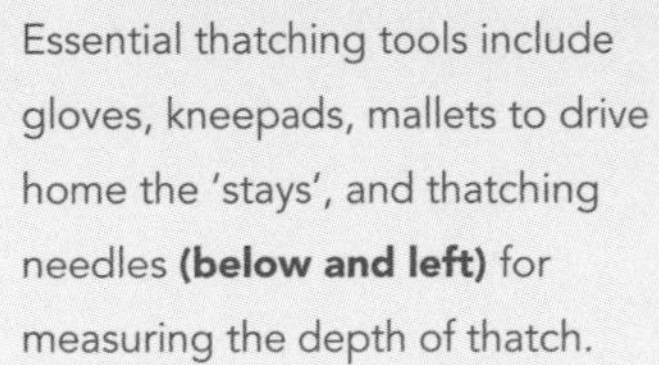

Above: The finished thatch is three times as thick as the original layer due to the compression.

Left: The original layer can be seen on the inside, near the wall.

Right: Notice the 'eyedrop' gable window and the way the thatch skirts the flat-roof extension at the back of the house. This thatcher is known for his windows, several of which have won awards.
Below: The ridge has a pretty hazel decorative feature that helps to keep the ridge in place.

Irish Thatch

Much of Irish thatch is in private hands
and functions as a family home.
Here are several examples around the
country that are open to the public.

LEINSTER

1. Irish National Heritage Park, pp12-9
2. Kilmore Quay, pp58-9
3. Tacumshane Mill, p69
4. Westcourt, p60
5. Michael Dwyer's Cottage, pp66-7
6. Abbeyview Cottage, p81
7. Treacy's Bar, p81
8. Skerries Mills, p69
9. Man O' War pub, p72
10. McDonough's Pub, p79
11. Foley's Tea Rooms, p75

MUNSTER

12. Dunmore East, pp84-7
13. Bóthan Scoir, pp88-9
14. Swiss Cottage, pp94-5
15. Adare, pp96-9
16. Bridge House and Cottage Pub, p102
17. Thatch & Thyme, p105
18. Deenagh Lodge, pp110-1

CONNACHT

19. Patrick Pearse's Cottage, pp140-1
20. Inis Mór Cottage, p131
21. The Man of Aran B&B, p133
22. Tigh Nan Phaidi Cafe, p135
23. Thoor Ballylee, pp124-5
24. Paddy Burke's Pub, p120
25. Moran's of The Weir p121
26. McDonagh's Pub, p114
27. Cong, pp146-7
28. Elphin Windmill, p69

ULSTER

29. The Thatch, p163
30. Glencolmcille Folk Village, pp148-59
31. Jeyffry's House, pp164-5
32. Derrymore House, p170
33. Ulster Folk and Transport Museum, pp171-5
34. Groomsport, p176
35. Andrew Jackson Cottage, p178
36. Arthur Cottage, p181
37. The Thatch Inn, p180
38. Dan Winter's house, p168
39. Crosskeys Inn, p177
40. Hezlett House, p184
41. Wilson Ancestral Home, p187
42. Coyles' Fisherman's Cottage, p188
43. Sheelin Lace Museum and Tea Shop, pp166-7

Northern
Ireland
ULSTER
CONNACHT
LEINSTER
MUNSTER
Donegal
Derry
Antrim
Tyrone
Fermanagh
Armagh
Down
Monaghan
Sligo
Leitrim
Cavan
Louth
Mayo
Roscommon
Longford
Meath
Westmeath
Dublin
DUBLIN
Galway
Offaly
Kildare
Laois
Wicklow
Clare
Carlow
Kilkenny
Tipperary
Limerick
Wexford
Kerry
Cork
Waterford
1
2
3
4
5
6
7
8
9
10
11
12
13
14
15
16
17
18
19
20
21
22
23
24
25
26
27
28
29
30
31
32
33
34
35
36
37
38
39
40
41
42
43

THATCH THROUGHOUT THE LAND

The local credit union in Knockananna, County Wicklow. **Opposite top:** Ripe barley in the background, unripened oats in the foreground.

LEINSTER – GO EAST AND SOUTH

The most populated of Ireland's four provinces, and home of the capital, Dublin, Leinster comprises twelve counties: Carlow, Dublin, Kildare, Kilkenny, Laois, Longford, Louth, Meath, Offaly, Westmeath, Wexford and Wicklow.

Wexford and Kilkenny largely make up the 'sunny south-east' and are home to several mud-walled, usually scolloped or thrust-thatched homes. From the Wexford coastal fishing village of Kilmore Quay to villages such as Licketstown and Glengrant in south Kilkenny, fine examples of thatched properties are to be seen.

Thatch is also to be found in the fertile land of the Boyne valley, in County Meath, an area traditionally rich in water reed. The Boyne is steeped in history and folklore, from the ancient remains of Newgrange to one of the most divisive battles in Irish history — the Battle of the Boyne.

There is thatch aplenty around the towns and villages of north County Dublin, an area of rich arable land.

Reaching further north into County Louth the climate suits the cultivation of corn, its straw being a good source for thatch.

In the province, thatched buildings can also serve a public function. Whether they are museums (the rebel Michael Dwyer's mountain hideaway) or the local credit union (Knockananna, County Wicklow) or montessori school (Skerries, County Dublin), they are a rich addition to our built environment.

Blackwater Barbers
Elaine
087 9514740
Sorry
WE'RE
CLOSED

Helped by the climate of the 'sunny south-east', thatch is in rich supply in County Wexford, from the huge variety of thatched skill on display at the Irish National Heritage Park at Ferrycarrig to the pretty village of Kilmore Quay.

Opposite: Blackwater is near the sea with a beach at Ballyconnigar, County Wexford. A close-up reveals the clever details of the shells within the mortar.
Top: A charming L-shaped home in Ballineskar, County Wexford, with quirky, uneven-shaped windows and slanting roof.
Right: Kilmore, County Wexford, an example of reed thatch.

Above: The only surviving thatched house in the village of Kilmuckridge, County Wexford. The exact date of the house is not known, but it appears on an 1841 map in a cluster of houses with the church and a national school.

The first known inhabitants of the house were Patrick and Mary Redmond, both national school teachers in the 1860s, and the great-great-grandparents of the current owners.

Right: Beautifully detailed ridgework on a restored building, Cahore, County Wexford.

Above and right: A new roof of golden thatch, near Curracloe, County Wexford, with an unusual window over the front door.

Left: Newly reed-thatched cottage at Whitegap, Curracloe, County Wexford. This tiny cottage has windows that are no more than a foot high.

Left and above: A cottage near Curracloe, County Wexford, with lovely additions of skillets and lamps from a bygone age – all set off with a beautiful creeping rose.

Below: A lovely home at Kilmacoe, County Wexford, flanked by some cordyline trees.

A beautifully carved nameplate 'The Dunes' announces this lovely cottage, which is hidden away, a stone's throw from the beach at Curracloe, County Wexford.

The Mayglass house

Above and left: Mayglass farmstead, near Bridgetown, in County Wexford, was built in the early eighteenth century. This mud-built home, with its associated outbuildings, retains many of its original features such as the hearth canopy and wall oven. Numerous artefacts and pieces of furniture also recall life in Ireland since the beginning of the 1700s.

Opposite: Traditionally, chimney breasts were constructed of masonry or wattle fabrications of clay plastered onto a wicker framework, and supported above the fireplace opening on large timber lintels called bressumers.

With a Mexican hacienda feel, the condition of this thatched house is extraordinary, given its proximity to the salty sea.

From 1921–2003, this Wexford thatched home in Cullenstown was completely decorated with shells by Kevin fFrench. The mussel, cockle, scallop and razor shells tell a story. In February 1914 the *Mexico* ran aground during a storm on the Keeragh Islands (which this house overlooks). The Fethard lifeboat, the *Helen Blake,* came to the rescue; nine of the fourteen lifeboatmen lost their lives. Tuskar Rock Lighthouse, just west of Cullenstown, is also illustrated on the house. Tuskar also witnessed tragedy when in 1968 an Aer Lingus plane crashed into the sea with the loss of all on board.

A beautiful home overlooking the Saltee Islands, near St Patrick's Bridge. Kilmore Quay, County Wexford.

Below: The main street in Kilmore Quay. Note the straw dolly **(inset)** on the right-hand edge of the ridge.

Kilmore Quay is a working fishing port in south County Wexford. This beautiful village remains largely unspoilt and still has many unique customs and elements of folklore that have survived from ancient times. Up until the nineteenth century 'Yola', a dialect of Middle or Old English, was still spoken here.

One of the most impressive visual aspects of the village is the main street **(above)**. Here the mud walls are constructed with 'marl', clay particular to the area, which is combined with seashells and chopped straw.

County Kilkenny retains quite a number of thatched cottages. The Barrow, Nore and Suir rivers traditionally supplied a wealth of reed to maintain these buildings.

Above: Westcourt, Callan, County Kilkenny, is the birthplace of Edmund Rice, the founder of the Christian Brothers, and it is preserved as a museum. The six-roomed house is almost three hundred years old. Rice was born in 1762 when the Penal Laws against Catholics were still in force in Ireland. County Kilkenny fared better than many places because of the tolerance of John Butler, the Protestant Duke of Ormonde who resided in Kilkenny Castle.

Left: Clogh is home to the last mining cottage in County Kilkenny. This small village lies north of Castlecomer, a large coal mining area from the 1600s right up until 1969. Phil Barron's cottage was built in 1890 and still has the original windows, door and furniture, including a settle bed made from a local ash tree. Notice the buttress supporting the front wall and the unusual position of the back chimney on the side of the roof.

Above: A charming house at Kell's Priory, situated in a row very near the mill race, on the River Kings, a tributary of the Nore. The thatch carefully lips over the porch.

Right: An attractive house in the Kilnamanagh area with unusual windows on the second floor. The magenta accents work well with the whitewash.

Some limewashed and brightly painted cottages in the area of Licketstown, close to the River Suir. Licketstown, Glengrant and Corluddy in Kilkenny are well preserved 'clachans' (from the Irish '*clochán*', meaning 'stone cell'). Dating back to medieval times, these settlements are often informal clusters of farm or fishing buildings, lacking a church, public house or other formal building, but still forming a type of village. Ownership boundaries are blurred, and the land connected with the farm buildings may be some distance away. The clachans close to the Nore and Suir rivers made good use of the local water reed. These settlements were virtually wiped out as a result of the Great Famine (1845-49).

Above: These buildings at Licketstown show the integrated nature of the farm buildings, grouped together in an informal manner creating a distinctive sense of place, separate from the operating farm.

Right: Glengrant is another area known for its surviving clachan dwelling.

Above and left: The above farm at Luffany comprises the house and old thatched dairy and other outbuildings, in the form of a clachan. At one time the area had many farms like this. All the farms had some fields near the village and others at a greater distance. Most farms were small with a few cows, a ploughing horse, pigs and sheep.

Above and right: Many of the Luffany thatched houses incorporate porches, buttresses and lovely shuttered windows. There are different thatching terms in different regions, here the ridge is known as 'breac'.

County Wicklow has a number of thatched properties. One of the best known is Michael Dwyer's cottage, at Derrynamuc, which is a museum looked after by the state.

Above: Michael Dwyer was a United Irishman leader in the 1798 rebellion. He led a guerilla campagn against local loyalists and yeomanry. Betrayed by an informer, Dwyer went into hiding in the Wicklow mountains in 1799. Surrounded by British troops, Dwyer managed to escape. Eventually, the authorities apprehended him and he was transported to New South Wales, Australia. A colourful character, he spent time in Van Diemen's Land (Tasmania) and was also chief of police in Liverpool, New South Wales. Dismissed for drunken conduct, and falling on hard times, he ended his days in Sydney's debtors' prison, where he contracted dysentry and died.

His Wicklow hideaway, made from local stone, lay in ruins for many years, but was restored in the late 1940s and repaired and rethatched again in the 1990s. The 'scraws', or fitted sods, sit directly on the roof timbers **(see inset opposite)**. The scraws help keep out the cold and damp and are a good source of anchorage for the 'scollops', or 'stays', which secure the thatch.

Opposite: Inside the restored cottage is a beaten clay floor, open wattled fireplace with a crane (for cooking vessels), teapot and skillet and some examples of country furniture.

County Dublin is associated with the urban capital, but there are a number of thatched properties throughout the county, especially in the coastal villages of Rush, Lusk, and Skerries.

Above: An attractive and colourful streetscape in Skerries.

Below left: Intricate ridgework forms a place for the birds. **Below right:** The purple woodwork stands out.

Right: Skerries has two restored mills, and this one, dating from around 1525, is still thatched. Flour has been milled here from the twelfth century up until the early twentieth century. It is thought that windmills came to Ireland with the returning Crusaders. The Skerries Mills are both 'tower mills', where the cap alone turns to face the wind. The thatched wooden cap is turned to the wind from inside the mill by hand-winch. When the wind blows and the sails turn, power is transmitted downwards via the central shaft to a single pair of grinding stones. Grain is carried manually to the top of the mill.

There is a thatched windmill at Tacumshane, County Wexford, and Elphin Windmill, County Roscommon, is also of thatch.

Left: A lovely thatched montessori school in Skerries.

Above and left: Lizzie's Cottage, an unusual L-shaped cottage near Loughshinny, County Dublin. The decorative work is not only on the ridge, but also works its way down to the base of the roof.

Opposite main: Casino House in Malahide, County Dublin, was once a substantial and prosperous dwelling, it is typical of late seventeenth-century architecture with the wide door, overhead fan window, and the floor to ceiling large windows. There was some discussion about it becoming the home of the Fry Model Railway.

Opposite inset: This lovely little house is hidden away in the busy suburb of Raheny, Dublin 5.

The Man O' War pub near Balbriggan is worth a visit, not only because of its appearance and character, but also for its kitchen. Buildings like these have character and charm, and provide plenty of work for talented craftsmen.

County Louth is home to some very pretty thatch, in particular near the coast.

Above: The colourful main street of Clogherhead. **Below:** The houses of the wonderfully named Crooked Street peep out at the road (the one on the right has unusual eyebrow dormer windows).

Above and left: Baltray (or *Baile Trá*, meaning 'beach settlement') is a small village on the north of the Boyne estuary, close to Drogheda, in County Louth. The mud flats are home to little terns, and the shoreline is dotted with well-preserved thatched houses, including this one with rich brickwork.

Below: Pretty cottages looking out across the estuary.

Above: With a well-integrated porch, thick ridge and lovely brick chimney stack, this property in the Carstown area stands out.
Right and below: Foley's Tea Rooms in Castlebellingham has a number of interesting features, not least the red-brick, Tudor-style chimney stacks. The brightly painted door, traditional shopfront, shop signs, flower boxes and other accoutrements bring colour on the greyest of days.

Above and inset: A fine stone and brick, whitewashed house from about 1820, being rethatched with golden straw at Kilcroney, County Louth. The site of the last eviction from County Louth, a nearby plaque reads:

> *Erected by Louth farmers in memory of Laurence Crawley whose eviction from this house by Lord Louth on 11th June 1881 was a turning point in land reform.*

Right: In this whitewashed house in Tully, County Louth, the tradition, common to places like Donegal, of the 'small house' in the garden of the big house can be seen.

Right: Golden thatch adorns this former public house, built around 1850 at Newtownstalaban, near Drogheda. The corrugated iron dormers are unusual and attractive.
Below: This considerable property in Ramparts, near Carlingford, is set back off the road with many farm outbuildings. The detailed work on the wooden door and surround is particularly interesting.

County Meath is home to the Boyne river, a traditional source of roofing material, and has its own thatched heritage. This house has a particulary dramatic aspect as it reaches right down to the sea at Laytown.

Opposite top left and top right: Thatched porch and McDonough's pub, Bettystown, with a row of thatched cottages looking out to sea **(opposite below)**.

Heineken Lager
Heineken
Premium Quality
BAR
PUSH

Inland in Leinster, County Laois has a number of surviving thatched homes.

Above: Golden straw is set off with a brightly painted green door at Jamestown, near Ballybrittas.

Left: A building in attractive decay at Cloonagh, near Mountmellick.

Above: Treacy's Bar, The Heath, Portlaoise, offers a welcome respite from the road. Whether tea-room or inn, the many thatched hostelries throughout the country owe part of their staying power to their distinctive roof and old-world features, helping them to stand out as landmarks, and give a welcoming feel.

Right: Self-catering Abbeyview Cottage, which may even date as far back as *c*1468, at Jamestown has many unusual features, including the windows by the door and the large sash windows.

MUNSTER – WAY DOWN SOUTH

Ireland's largest province, Munster stretches from the lush fields of County Waterford in the south-east to the wilder coasts of Cork and Kerry in the south-west and includes Tipperary, Limerick and Clare. The pretty, largely thatched village of Dunmore East is a thriving fishing port and a highlight in terms of thatch in east Waterford.

With the rivers Nore and Suir traditionally providing water reed, Tipperary boasts many thatched properties, from the simple peasant dwelling of Bóthan Scoir in Cashel to the elaborate '*cottage ornée*' at Swiss Cottage near Cahir.

Limerick also has quite a number of thatched houses, but perhaps none so spectacular as those at the heritage village of Adare, built by the Earl of Desmond in the early nineteenth century.

Travelling through places like Kanturk, Boherboy, Charleville, and Mitchelstown, in Cork, Ireland's largest county, there are many thatched buildings to catch the eye. Thatch is also a feature of the rich pastureland known as the Golden Vale, bordered by the Galtee Mountains, the Glen of Aherlow and the Blackwater river valley. The area takes in parts of counties Tipperary, Limerick and Cork.

Kerry is home to the famous lakes and mountains of Killarney National Park as well as miles of stunning rugged coastline. Thatched buildings such as Deenagh Lodge fit perfectly into this wild natural environment.

There are plenty of thatched buildings scattered throughout County Waterford. A great many are clustered together in the seaside village of Dunmore East.

Above and right: This house on an entrance road into Dunmore East has a fantastic snake-like ridge.

Dunmore East is a lovely fishing village on the western side of Waterford harbour, facing Hook Head. There is a plethora of thatched buildings, and some of the contemporary village homes are sympathetic to this environment by also having thatch.

Above: A row of thatch greets the visitor on entering Dunmore East.
Left: This house outside Dunmore East village has an eye-catching hipped roof, with its door in an unusual position.

Above: One of the thatched homes in a row in Dunmore East.
Right: Pretty thatch and flora in the village.

County Tipperary is rich in a variety of architectural buildings, in particular those of an ecclesiastical nature. The Rock of Cashel was a great powerhouse of early Christianity. The county also has its fair share of surviving vernacular housing.

Left and below: The Bóthan Scoir (peasant dwelling), just outside Cashel, is a preserved tenant cottage, dating from around 1640.

The Bóthan Scoir is the last of ten small farm labourers' dwellings that once stood near Cashel. Four of them were built of stone, similar to this one, and had thatched roofs; the rest were made of mud and wattle. The labourer's cottage was usually the house of a migrant worker, and was often abandoned once the worker moved on. Opposite the Bóthan is the 'ducking pond', where unpleasant punishment was dished out to ladies who were guilty of socially unacceptable behaviour. Notice the fireplace without a chimney **(left)**, and the cooking implements on a crane.

Below left: Many thatched museums hold nationalist memorabilia.

Below right: Spare reed thatch hides behind the bright red half-door.

Above: Dating from c1800, the house above in Boherboy, County Tipperary, displays much of its original character, retaining the older windows, and sporting a very fine thatch with sturdy ridge. **Left:** A tastefully modernised thatched house at Ballinlough. Notice the stone plinth over the door.

Ballycahill has a few thatched houses left, with a welcome sign which is also thatched. Sadly, The Thatch public house has seen better days.

This two-storey house, which is surrounded by tidy vegetation, has a pleasing aspect in the village of Cloneen, County Tipperary.

Above and inset: Just on a bend in a road in rural southern Tipperary, this beautifully renovated house at Mohober stands out.

Left: A property near Holycross with unusual windows on both the slate and the thatched house.

Opposite: A farm complex at Lawlesstown, just outside Clonmel.

Swiss Cottage, at Kilcommon, near Cahir in County Tipperary, is a fine example of a '*cottage ornée*', or ornamental cottage, built in the early 1800s by Richard Butler, Earl of Glengall, to a design by the famous Regency architect, John Nash. It was originally part of the estate of Lord and Lady Cahir and used for entertaining guests. Whilst the idea 'played' with the idea of simple rustic life, the building nevertheless has some fine features. The balcony supports of local timber resemble growing trees, and the painted black linework is a natural extension of this. The balcony is roofed with wooden slates, which keep it rooted in the envirnment. The rolling pitch of the thatched roof also makes for soft integration with the landscape.

Opposite top: The timber supports, designed to be tree-like.

Opposite bottom: The interior of Swiss Cottage contains a graceful spiral staircase and some elegantly decorated rooms. The wallpaper in the salon, manufactured by the Dufour factory, is one of the first commercially produced Parisian wallpapers. This is a far cry from the traditional labourer's cottage, but is nevertheless a part of our rich thatched heritage.

County Limerick has the Shannon river running through it, and is home to a number of reed-thatched properties. Some are to be found south of the city of Limerick as far as Kildimo and reaching down to the Golden Vale. The picturesque village of Adare is particularly rich in thatch. These properties are also interesting in that they are much more in the '*ornée*' or even English-picturesque tradition.

The park in Adare has a thatched bandstand.

Right: This building dates from about 1800 and features a wonderfully unique timber porch, with natural tree trunks for supports in the *ornée* tradition.

A designated heritage town, Adare was built as the village for the Dunraven Estate, in the early nineteenth century, by the Earl of Desmond. Today, many of the cottages serve as bustling businesses as the village is a popular tourist spot, but many of them are also private homes.

Gift n Time
Gifts for all occasions
ENTRANCE

Cork, the largest county in Ireland, also has a substantial number of thatched houses, particularly in the rich agricultural region of the Golden Vale.

Above: A lovely house with outbuildings at Ballydeloughy, near Mitchelstown, County Cork.
Right and opposite: Boherboy, west of Kanturk, has some lovely thatched buildings. The integrated porch of this building, with its stone lintel above the door, is an unusual feature.

This page: Bridge House and the adjacent Cottage Pub, Dromina, County Cork. Both have some lovely stonework and an interesting wooden feature on the gable end. **Opposite:** A beautiful and impressive thatched porch at Coolclough, County Cork.

The entrance, window shutters and simple, but elegant, pillars of a fine property in Farahy, near Mitchelstown, County Cork.

Above: A striking brick chimney stack at Ballindangan, near Mitchelstown, County Cork. **Right:** 'Thatch & Thyme' at Kildorrery, County Cork.

Above: Kiltoohig, near Charleville, County Cork.

Below: Many thatched properties **(below and right)** still exist in the lush farmland of the Golden Vale, on the southern side of the Galtee Mountains, in such towns such as Charleville, Mitchelstown, Kilmallock and Tipperary.

The church at Newtownshandrum, near Charleville, County Cork, has an elaborately 'thatched' noticeboard, alluding to the craft in the area.

Lovely thatched home with matching painted chimney stack at Newtown, near Charleville, County Cork.

County Kerry has few surviving old thatched cottages. However, a beautiful example of thatch is to be found at the popular tourist destination of Deenagh Lodge at the entrance to Killarney National Park.

Above: A glimpse of the spectacular surrounding parkland of mountains and river at the ornamental gatehouse.

Right: Deenagh Lodge dates back to 1834 and was the gate lodge for the Kenmare Estate. Among its many fine features are the magnificent, tall, octagonal chimney stacks, lattice glazing and external roof supports. Its Gothic-revival look is more in keeping with a *cottage ornée* than a working man's dwelling. Today it is a tearoom.

CONNACHT – INTO THE WEST

The wild, rugged landscape of the Irish west provides a spectacular backdrop for thatch. The five counties of Connacht – Galway, Mayo, Leitrim, Roscommon and Sligo – all bear testament to this heritage. From east Galway around Lough Corrib to the savage beauty of Connemara, where Patrick Pearse's cottage nestles into the hills around Rosmuc, and from the wildly exposed Aran Islands, battered and defined by the Atlantic Ocean, to the lush area around Cong in Mayo, the variety of landscape is startling. On a fine day there is surely no better place. It is the least-inhabited province of Ireland and still retains a rich Gaelic heritage with Irish language communities in the Gaeltacht of Connemara and Aran. Galway is the largest urban centre in the area, and it too has a rich heritage of thatch in the nearby village of Menlough.

In this region there is a noticeable change in the type of thatch, as the weather and Atlantic winds demand a different approach. Here we find surviving examples of roofs thatched with rye, which traditionally grew in the area, held in place with ropes, hence the term 'roped thatch'. Part of this tradition are the *bacáns*, stones or rods that hold the ropes in place either at the gable or the eaves. Another notable feature is the crow-stepped gable, giving many a pretty whitewashed cottage a crenallated appearance. Limestone walls are not only evident as boundary dividers, but are also used to construct the houses.

The wild western seaboard of County Galway and Connemara is home to a large number of surviving thatched properties. The area around Lough Corrib, stretching north to Mayo also has a rich supply.

Left and below: McDonagh's pub on the main street of Oranmore, County Galway, sports a fine reed-thatched roof.

Above and right: This house at Oranhill, near Oranmore, has unusual features, such as the large convex window and door surrounds. The Marian shrine matches up with the colour code.

Above: Two fine houses near Oranmore, the one on the left peeping out from ample foliage, the one on the right with its unusual window set into the thatch above fine stonework.

Below and inset: House and outhouse in Ballynacourty near Kinvarra with matching flowers and painted sills.

Above and right:
Two houses, one at Carrowmore, south of Oranmore, and one west of Clarinbridge on an inlet of Galway Bay.

Above: Carrowmore, south of Oranmore, looking towards the sea estuary of Galway Bay.

Left: Two sliotars and religious stained-glass pieces fit the scene.

Right and below: Carrowmore, with its stone walls, south of Oranmore, on the way to the sea estuary of Dunbulcaun Bay, the easternmost part of Galway Bay.

Left and below: Clarinbridge, meaning 'small board bridge', is known for its oyster festival which has taken place every September since 1954, and attracts visitors from all over the world.

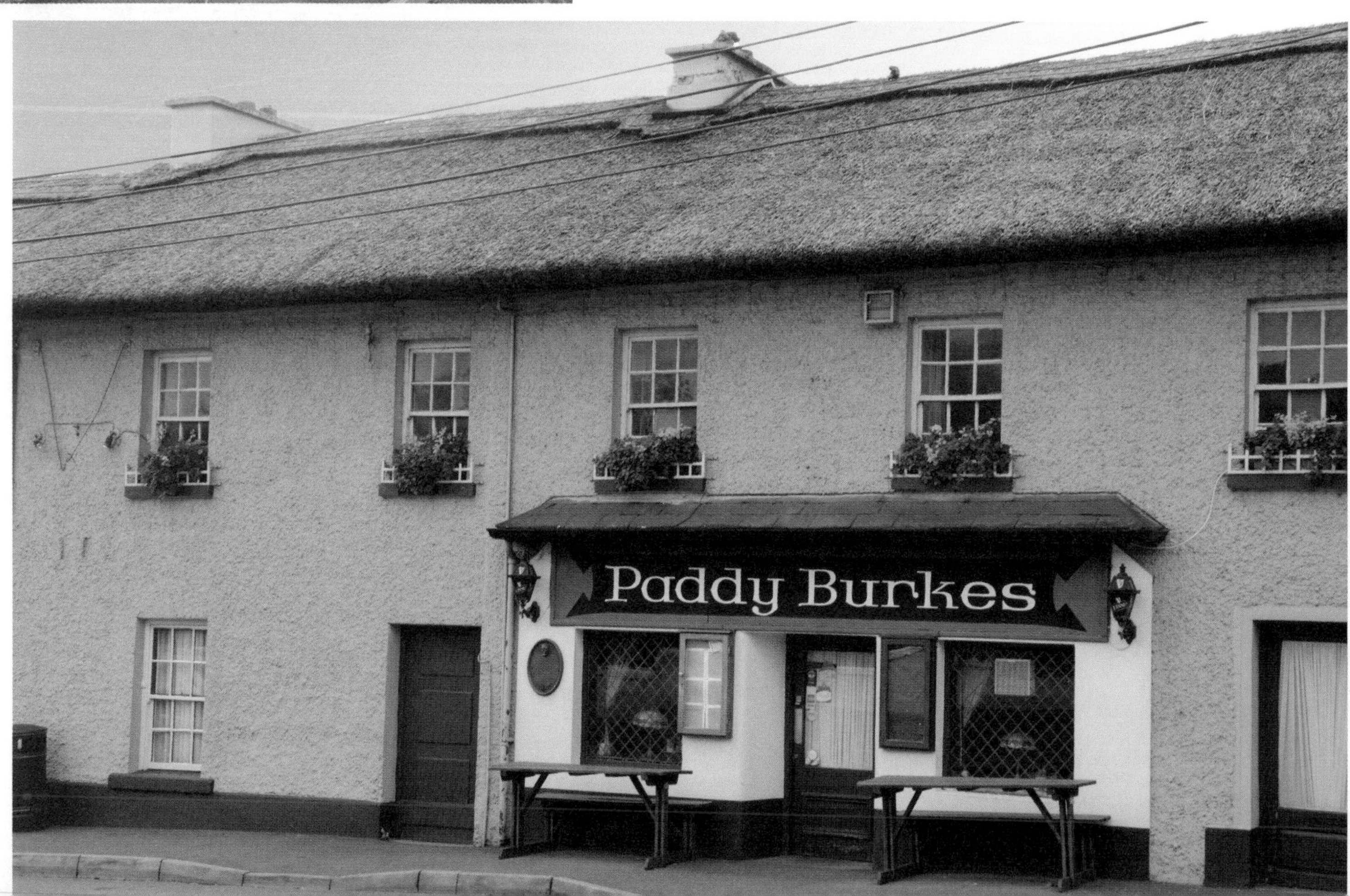

Above and below: The Weir at Kilcolgan is a former fishing village on an inlet of Galway Bay that is home to a number of thatched properties, including the well-known Moran's, world-famous for its oysters. It is currently run by Catherine Moran, the seventh generation of the family in the business.

Above: A gorgeous cottage south of Galway city, which looks as though it has had the end sliced off, but is charming nonetheless.
Below: Driving cattle past a lovely thatch near Kinvarra.

Left: Some new thatched buildings across the road from the impressive **(above)** Dunguaire Castle, Kinvarra. **Below:** The picturesque fishing village of Kinvarra has its own fair share of thatch.

Thoor Ballylee castle, near Gort, south County Galway, is a sixteenth-century tower-house built by the de Burgos. It is known as Yeats's Tower as it was home to the poet William Butler Yeats from 1921 to 1929. In the early 1900s the castle was part of the nearby Coole Estate, home of Lady Augusta Gregory, Yeats's lifelong friend. Coole House hosted many Irish Literary Revival gatherings.The tower is now in the hands of the state, and the adjoining thatched cottage is a Yeats museum. A plaque on the building reads:

I, the poet William Yeats,
With old millboards and
sea-green slates,
And smithy work from
the Gort forge,
Restored this tower
for my wife George.
And may these characters
remain
When all is ruin once again.

Above: Near Annaghdown and Corrandulla, County Galway. Annaghdown is an eastern inlet of Lough Corrib, so a rich area for water reed.
Below: Many of the homes in this area feature detailing around the windows and doors and just under the roof.

A thatching in progress. You can see the uncut reeds on the ridge and the difference in colour between the old and new straw.

Above: The building above in Annagh, County Galway, has an unusual attic window and exposed stonework. The wicker-work in the garden marries perfectly with the cottage.

Above: With one half a thatched dormer and the other a two-storey dwelling, this pretty house in Balrobuck Beg, south of Headford, County Galway, has several eye-catching features, including the use of red on the windows and shutters. **Below:** A cosy building is tucked in nearby in Balrobuck More.

A lovely building in a fabulous setting, Bunatober, County Galway, again with the regional details of decorative render surrounds.

Below: A very well-maintained home and garden near Headford.

As the traveller leaves the mainland at Ros a Mhíl towards the Aran Islands, thatched properties nestle by the sea.

The Aran Islands is made up of three islands: Inishmore (Inis Mór), Inishmaan (Inis Meáin), and Inisheer (Inis Oírr). At one point, most of the homes on the Aran Islands would have been thatched. Today, there are only a few left, but they offer stunning views and stand as a testament to those who preserve them.

This lovingly renovated building is available for rent on Inis Mór. Notice the small model of the house in the garden.

Above: The remains of a thatch on Inis Mór are beautiful in their own right as nature slowly reclaims the building, and the house gazes across to the mainland **(below)**. Notice how the thatch was tied down **(left)** against the Atlantic gales.

Above: The Man of Aran B&B on Inis Mór. The island buildings are very exposed as there are few trees; the only real barriers to nature are the limestone walls, which are used to mark boundaries. **Below:** A thatched home nestling into the landscape.

The crow-stepped gables are typical of Atlantic coast and island thatched houses.

Above: Tigh Nan Phaidi Café, Kilmurvey, Inis Mór, is a great place for replenishment before you face up to the walk towards Dún Aonghasa. Its thatched roof adds to the charm.

Right: Just across the road is the Kilmurvey Craft Village, selling all things Aran. Tourism is an important source of income for these islands.

A fine slate-roofed cottage with thatched outhouse. The Galway mainland is visible in the distance.

Many of the buildings on Aran were thatched with rye straw.

Looking across Galway Bay to the Twelve Bens in Connemara.

Roped rye-straw thatch weathers to a wonderful silver colour, so that even in decay the thatch has visual appeal.

Above: Farmers moving cattle past a ruined former thatch on Inis Mór. You can still make out the crow-stepped gable.

Opposite: The edge of the world – the dramatic view out to sea from Dún Aonghasa shows just how exposed Aran is. It is no wonder there is little thatch left.

The dramatic environment of Patrick Pearse's cottage, in Inbhear, near Rosmuc in Connemara.

This small restored cottage, overlooking the breathtaking lakes and mountains of Connemara, was used by Patrick Pearse (1879–1916), one of the leaders of the 1916 Rising, as a summer residence and summer school for his pupils from St Enda's, in Dublin. The cottage was burned during the War of Independence, but has been restored and is now in the care of the state.

Above: This fine house with different levels for the dwelling house and outhouse is at Killannin, near Oughterard, County Galway.

Below: Oughterard, County Galway.

Above: A stunning cottage with floral display, Spiddal, County Galway. **Right and below:** Connemara thatch.

Above: This house peeping through the trees in Menlough, just outside Galway city, has distinctive colouring and windows.

Above: On a bend on the road, Menlough, County Galway.
Right: This building has unusually small windows, so there would be very little natural light inside.

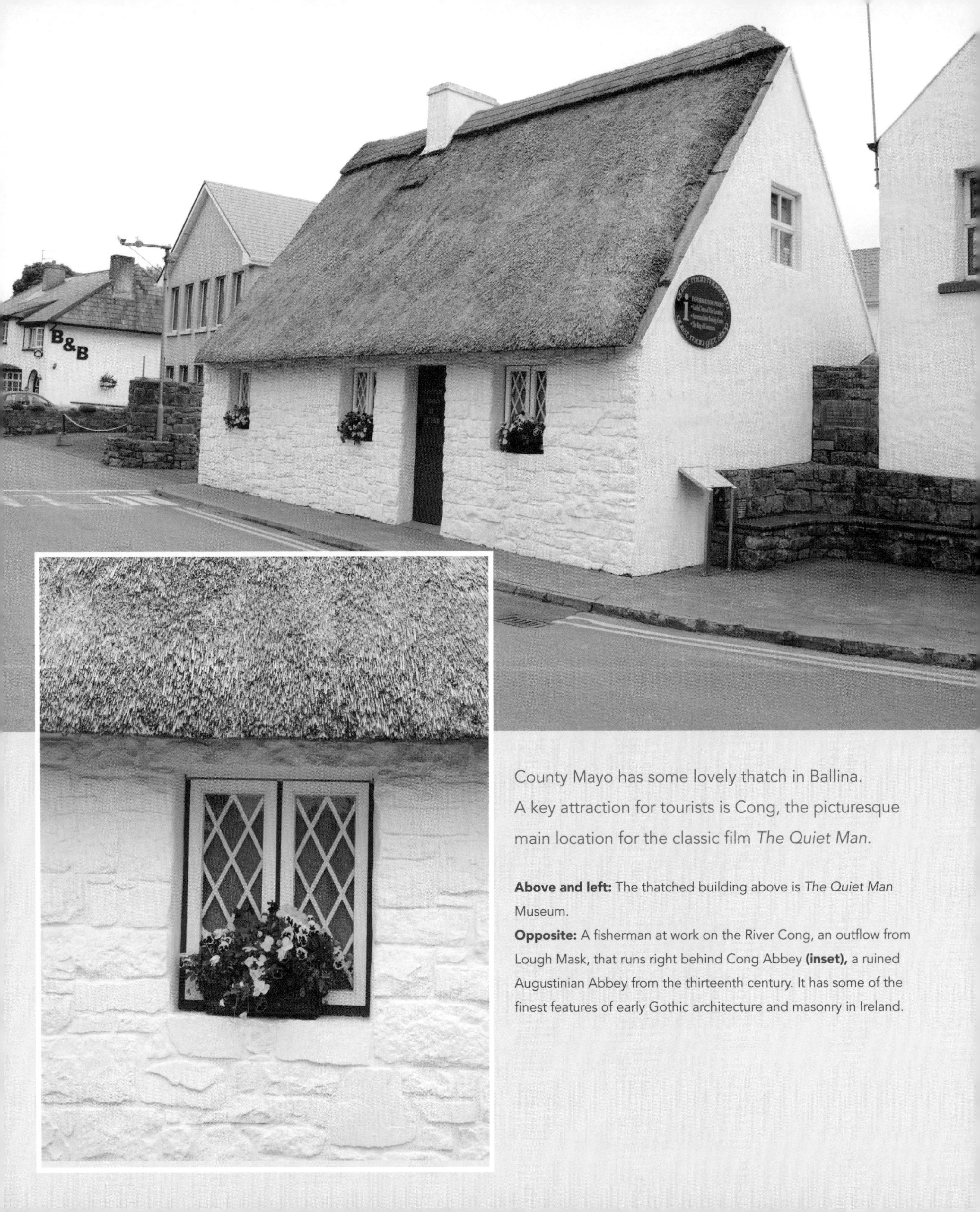

County Mayo has some lovely thatch in Ballina. A key attraction for tourists is Cong, the picturesque main location for the classic film *The Quiet Man*.

Above and left: The thatched building above is *The Quiet Man* Museum.

Opposite: A fisherman at work on the River Cong, an outflow from Lough Mask, that runs right behind Cong Abbey **(inset),** a ruined Augustinian Abbey from the thirteenth century. It has some of the finest features of early Gothic architecture and masonry in Ireland.

A fine thatch nestled in the stunning and dramatic shadow of Slieve Tooey in the environs of Glencolmcille, County Donegal. Notice the new bungalow **(inset)** built a little further down the road.

ULSTER – THE NORTHERN REACHES

Ulster's geographical area lies in the north of Ireland, from the wild Atlantic coast of County Donegal to Antrim, Armagh, Derry, Down, Fermanagh, Monaghan, Cavan and Tyrone. In the six counties of Northern Ireland (Antrim, Armagh, Derry, Down, Fermanagh and Tyrone) there are about one hundred and fifty thatched properties; in the rest of Ireland there are around three thousand.

County Donegal boasts its own unique version of the craft, and there are several surviving examples, as well as some reconstructions, such as those at Glencolmcille. Fr McDyer's Folk Village Museum at Glencolmcille was founded as an initiative to create employment and opportunity in a community devastated by unemployment. The buildings are roofed in the traditional Donegal way, employing a traditional 'A-frame', or rounded roof, with roped thatch tied on with bacáns.

Within the six counties of Northern Ireland, there are rich examples and different varieties of thatch. From the 'cruck-frame' houses of the Plantation in the seventeenth century, to one-room fisherman's cottages. Several thatched properties are in the hands of the National Trust and open to the public. There are also a few ancestral homes of American presidents dotted around the region. Flax, of huge economic importance to the North in the past, was traditionally used for thatch. There are several examples in the wonderful Ulster Folk and Transport Museum.

A wild and windswept county, Donegal also bears witness to its own unique version of Ireland's surviving thatched heritage.

The Folk Village Museum, at Glencolmcille, was the initiative of dynamic Donegal priest Fr James McDyer. It opened in 1967 with three cottages, built and furnished in just three months through local community effort. At the time of Fr McDyer's arrival in 1951 the parish of Glencolmcille was suffering from a long-standing cycle of unemployment and emigration. Over the next three decades Fr McDyer worked to help break this cycle, lobbying for amenities, organising community projects and supporting the development of local industries, many of which, like the Folk Village, are still successful today.

New cottages and exhibits have been added over the years, and the Folk Village Museum is still very much a community concern. Local people donate time, expertise, artefacts, stories and photographs. Tucked into a rocky hillside, the Folk Village offers visitors a warm Donegal welcome and an intimate experience of a past way of life. The buildings explore life as it was in the 1700s, 1800s and 1900s. The thatched cottages are exact replicas and are furnished accordingly. The thatch has been maintained by Ivor Kilpatrick.

Above: The collection of houses, or clachan, at the folk park on a cold winter's day.

Left: Roped thatch.

Above and right: The thatch is held down by ropes, attached to bacáns, or stones **(see right)**, that project from the wall at regular intervals near the eaves. The ropes would have originally been *súgán* – hay or straw rope.

Left:
Replica interior of one of the cottages. Notice the A-frame and the curtained bed beside the hearth.

Top left and right: This replica of a fisherman's cottage shows all the accoutrements of his trade: seaweed, drying outside the door in a basket, probably for use on the land, and inside a collection of nets hanging from the ceiling and clothing drying over the fire.

Right: Historically, a clachan was a group of small single-storey cottages, usually belonging to farming or fishing people and sited on poor land. Each cottage in the Folk Village 'clachan' recreates a different era of Irish history. The cottages are all neatly white-washed, with a traditional half-door design to keep the animals out and the people in. Roofs are thatched in the distinctive rounded Donegal style, tied down securely with rope and pegs to protect the thatching from the fierce westerly winds off the ocean.

Left: A striking home, nestled into the hills near Glencolmcille.

Above and right: A beautiful restored house and outhouse in Glencolmcille, the single surviving building of a clachan. Notice the bacáns, or stone pegs, which were used to tie ropes over the thatch to keep it in place. One of the characteristics of vernacular building is the fact that everything is uneven. It results in real charm. The roof is typical for the area, low and rounded to protect it from the prevailing Atlantic winds. You can see some of the older thatch underneath.

Above: A lovely thatched house with bright yellow door and matching sills near Glencolmcille.

Below and inset: An Srath Buí in the environs of Glencolmcille commands a great view from its sloping site.

Above, inset and below: Two houses peep out of their respective landscapes near Glencolmcille. These houses blend into the landscape rather than fight it.

Above: An impressive avenue announces this house near Glencolmcille, County Donegal.

Left: Bridge Cottage, Casheltown. A direct-entry cottage near Killybegs.

Opposite top and inset: This cottage sits so well in its environment in Shalvey, west of Killybegs.

Opposite bottom: Stored turf, thatched with flax and roped, near Glencolmcille.

Cyndi Graham's handweaving studio is near the beautiful St John's Point, south County Donegal. It is interesting to see this home in use today, probably for the same purpose it was originally intended.

This cottage **(inset)**, adjacent to Ballysaggart Pier, near St John's Point, may have once served as a fisherman's cottage. Houses in this area have unusually large sash windows in comparison to other thatched cottages around the country.

Above: St John's Point.

Below: A tiny restored cottage in Mountcharles, County Donegal. A plaque on the building records that this is the last surviving inhabited dwelling from Tamhnach an tSalainn, or 'field of salt', the original village that existed prior to the establishment of the town of Mountcharles by the Conyngham family during the seventeenth century.

Violet Cottage,
Behy, near
Ballyshannon,
County Donegal.

Left: The Thatch, Dorrian's pub, Ballyshannon.

Right: A colourful cottage at Corporation, near Killybegs.

JEFFRY'S HOUSE

Designed and built by architect Thomas O'Brien and artist Emily Mannion, in 2014, in collaboration with local thatcher Ivor Kilpatrick and Donegal County Council, Jeffry's House, Ards Forest Park, Creeslough, County Donegal, is a folly made from thatch on a wooden frame. The structure is named after Jeffry's Lough, a nearby lake, which appears on older maps but has now disappeared. This is the first architectural commission in a forest in Ireland.

Standing at the edge of the forest, Jeffry's House offers shelter and views of the sea, sand dunes and mountains beyond. Emily Mannion explains the thinking behind the project:

'For us the structure had to set itself apart from the landscape whilst also being very much part of it. Using thatch helped to "soften" the geometric structure whilst also giving it tactile form. Thatch is very much a traditional craft that uses simple materials, and we were keen to keep the building materials as much part of nature where at all possible.

This particular shape posed a unique challenge for the thatcher. In many ways the project wouldn't have been thatched if it wasn't for Ivor's dedication and willingness to take on that challenge. Ivor suggested flax for the thatch as traditionally cottages in Donegal would be thatched with either flax or straw. When we went to visit Ivor on his farm he showed us the flax and we fell in love with the colour, the texture and the smell of it.

Myself and Tom worked on building the main timber part of the structure, and Ivor and his apprentice, James, did the thatching.'

Sensitive to the natural beauty of the landscape around it, Jeffry's House combines traditional thatching with a slender geometric form and is designed to allow nature reclaim the ground beneath.

The structure is open to the ground beneath it, and the idea is to allow the wilderness to grow back around and inside it.

Moving into the six counties of Northern Ireland, Fermanagh, home to Belleek Pottery, the town of Enniskillen and the stunning Lough Erne, is not without its fair share of thatch.

Above: A thatch with an interesting fan window near Carnagh Bay, County Fermanagh.
Right and opposite: Beautiful curves on this thatched lace museum and tea shop in Sheelin, Bellanaleck, County Fermanagh.

Grocer
DAN WINTER
Tobacconist
Spirit Merchant

STOP

THE FOCAL POINT
OF THE
BATTLE OF THE DIAMOND
ON
21st SEPTEMBER 1795,
WHICH LED TO
THE FORMATION
OF THE ORANGE ORDER
IN SLOAN'S HOUSE,
LOUGHGALL.

County Armagh has some noted areas of natural beauty, including, to the north, Lough Neagh and, in the south, The Ring of Gullion.

Opposite: This pretty cottage nestled into the road near Loughgall, County Armagh, is the ancestral home of Dan Winters, one of the founding members of the Orange Order. It was here after the Battle of the Diamond in 1795 that a group met to form the Orange Order. The house is still in the hands of the Winter family. **Above:** The sadly no longer functioning Mullaghbawn Folk Museum.
Below: Beautiful Camlough lake on the Ring of Gullion.

Above: Derrymore House, near Bessbrook, County Down, was the summer residence of Isaac Corry (1753–1813), Chancellor of the Irish Exchequer. The house, which was built in the eighteenth-century *cottage ornée* tradition, sits in over one hundred acres of woodland demense, and is in the care of the National Trust. It is open to the public in the summer. The stunning location offers views over nearby Newry and in the distance the Mourne Mountains. Among its pretty features are two floor-to-ceiling windows.

County Down, with its impressive coastline that leads to the bustling port of Belfast, is also home to the Ulster Folk and Transport Museum – a discovery park of life from a hundred years ago.

The Ulster Folk and Transport Museum is split into two separate museums: one dedicated to transport and the other a folk park uncovering traditions practised in the past. Many of the traditional buildings have been brought, stone by stone, from various locations across the north and rebuilt in situ. Through an interactive approach, the museum allows visitors experience many disappearing crafts, thus helping to keep them alive.

Soda farls, a traditional part of the Ulster diet, being cooked on a griddle over an open fire at The Old Rectory **(inset)**.

Left: A reconstructed forge at the museum. Notice the exposed scraw of the roof and all the blacksmith's tools hanging up. The blacksmith forged all manner of metal, including shoes for horses.

Below: A reconstructed weaver's house.

Below inset: a loom.

Opposite: Belfast was built on shipbuilding but also flax, very little of which is grown today. However, in the Victorian era it was a huge industry in Northern Ireland. Here is an example of it growing **(top left)**. It blossoms with a delicate blue flower that only lasts two days and is then picked and dried either in the sun **(bottom left)** or above the fire **(right)**. Its sinews are pulled apart and spun to be worked on a loom to make linen. Flax has also been used for thatching in the north of Ireland.

Above: A reconstructed hill farm and **(left)** a mill where flax seeds were ground.

Bottom left and below: Many thatched houses in Northern Ireland were constructed with an oak cruck truss **(as seen below left)**, which was commonly used in Britain and by the seventeenth-century Plantation of Ulster settlers. Notice also the wall construction of rope straw **(below)**. Nearly four hundred years old, these surviving cruck frames were originally sited at Ballyvollen, County Antrim, near the Glenavy river.

Above: A cottier's house with the main house in the background at the Ulster Folk and Transport Museum.

Below: A reconstructed farmhouse from Corradreenan West in County Fermanagh, now sited at the museum.

Above: It is thought that these fishermen's cottages at the wonderfully named Cockle Row, Groomsport, County Down, are about three hundred and fifty years old. They are the last reminder of the fishing traditions of old in this area. Interestingly, they are built at a right angle to the sea for protection. The cottage on the left has a wonderful chimney stack, on the gable end. They were inhabited up until the 1950s. Groomsport is a little port on the busy shipping channel into Belfast Lough **(below).**

County Antrim is the county at the extreme north-east of Ireland and is known for the Giant's Causeway and the Glens. It, too, is home to some lovely examples of thatched buildings.

Above and right: The Crosskeys Inn, Ardnaglass, Toome, claims to be the oldest thatched pub in Northern Ireland. Built around 1654, it is famous for Irish traditional music. It has served as a coach stop on the Derry to Belfast road, a post office, and as a shop that sold everything and anything. It is close to Lough Neagh, Ireland's largest lake, whose banks traditionally would have been a good source of reed for the roof.

Left: Just outside Carrickfergus, County Antrim, Andrew Jackson Cottage is a traditional Ulster Scots thatched farmhouse, built in the 1750s. The centre hosts an exhibition on the life and career of Andrew Jackson (1767–1845), the seventh president of the USA, whose parents emigrated from here in 1765. In the grounds of the centre is the US Rangers Centre, a museum dedicated to the men of the first battalion of the US Rangers that was activated in Carrickfergus in 1942.

Below: Across the road is 'Fools Haven', another fine thatched property that is about two hundred and fifty years old. Here the author Ruddock Millar, orphaned after the *Titanic* disaster, was raised by an aunt. His works include *Stirabout*, *When Johnny Comes Marching Home* and *The Land Girl*.

Above: The commanding view from Andrew Jackson's cottage, across the waters to Belfast. To the right is the silhouette of Carrickfergus Castle, made famous by the song. You can just make out the Harland and Wolff cranes, 'Samson' and 'Goliath', on the distincitive Belfast skyline.

Above: Broughshane near Ballymena, County Antrim, is known as the 'Garden Village of Ulster', and has won a plethora of tidy-town-type competitions. It is also home to the attractive 'The Thatch Inn'.

Opposite: Nearby in Cullybackey, just a few miles from Ballymena, is 'Arthur Cottage', the ancestral home of Chester A. Arthur, the twenty-first president of the USA; his grandparents lived here. Acting as an interpretive centre, it tells the story of this impressive man's life. Constructed from local stone and thatched with flax, it has commanding views from its position on Gourley's Hill. It is open to the public in the summer.

County Derry is noted for the historic, walled town at Derry, but also for some spectacular scenery, on the North Atlantic coast at Mussenden, and some interesting thatch in the environs of Downhill Estate.

The ruins of Mussenden Temple perch on a hundred-and-twenty-foot cliff-top. Built in 1785, it formed part of the Downhill Estate of Frederick Augustus Hervey, Bishop of Derry and Earl of Bristol. The temple was built as a summer library and is inspired by The Temple of Vesta in Tivoli, near Rome.

Inset below: The stunning view from the temple.

On the walls of Mussenden Temple an inscription from the Roman poet and philosopher Lucretius reads: *'Suave, mari magno turbantibus aequora ventis e terra magnum alterius spectare laborem.'* 'Tis pleasant, safely to behold from shore/The troubled sailor, and hear the tempests roar.'

Above and left: Open to the public, Hezlett House, Castlerock, County Derry, is one of the oldest thatched houses in Northern Ireland and is a built around a cruck-truss frame. Originally a rectory, its pretty features include the fanlights above the doors and the flax-thatched folly in the garden. The house is located on the edge of the spectacular Downhill Demense with nearby incredible views of Mussenden Temple, Downhill Strand and the Donegal coastline.

Above and below: The impressive ruins of Downhill House.
Left: Mussenden Temple.

Above and left: Further into the nearby forest this charming cottage called The Pretty Crafty Design Studio peeps out onto the road in the heart of Downhill Forest, opposite the entrance to the Mussenden Temple.

County Tyrone borders Lough Neagh and boasts the natural beauty of the spectacular Sperrin Mountains. The shores of the Lough are home to a lovely example of a thatched fisherman's cottage.

Near Strabane lies the ancestral home of America's twenty-eighth president, Woodrow Wilson. His grandfather, James Wilson, lived here **(right),** emigrating across the Atlantic in 1807, to find work in the newspaper business and become a politician, known as Judge Wilson.

Above: Coyle's Cottage is a traditional fisherman's thatched cottage very near the shores of Lough Neagh at Cookstown, County Tyrone. It is also the starting point of the Gort Moss Walk, a linear walk along old moss roads. Like many lone surviving thatched properties, it is a focal point in the community for all types of local groups and cultural activity.
Right: The stunning waters of Ireland's largest lough, Lough Neagh.

GLOSSARY

BACÁN
A stone or metal peg under the eaves of the house that roped thatch can be fixed to.

BASE (WEATHERING) COAT
A roughly thatched coat acting as a base for the final coat.

BOBBIN
Handful of long straw twisted at its centre and folded in two, with an eye at one end. Bobbins are strung on strong hazel rods and secured with hairpin scollops to the apex of the ridge.

COPING OR RIDGE
Capping of clay or mortar.

DRAWING (PULLING)
Process of separating long and short stems of crushed (threshed) straw by wetting and drawing by hand into bundles.

EAVES
The lower edge of a sloping roof, which overhangs the wall head.

HIPPED ROOF
A four-sided roof having sloping ends and sides.

LEGGET
A toothed, card-shaped wooden implement, with a handle, used to dress the ends of water reed or wheat straw bundles tightly into their fixings.

MORTAR
A mixture of a binder (eg lime, clay or cement), sand and water used to bind stones/bricks. Mortar can also be used to make flashings, flaunching, copings and cappings.

RIDGE
The apex of a double-pitched roof and the capping covering it.

RODDING
Rods, usually of hazel, used to secure thatch to the rafters of the roof.

SCOLLOP
From the Irish word *scolb* – horizontal and bent rods of hazel, willow, ash, briar, bog fir etc used to secure thatch to the roof structure. The bent rods are shaped like hairpins or staples with the ends pointed.

SCRAW
From the Irish word *scraith* – a strip of pared lea (grassed) sod laid on the roof structure from ridge to eaves to provide a foundation and anchorage for fixing thatch.

SLICE
Thrust method of thatching, typical of east Leinster, in which handfuls of straw are thrust into the old thatch with a fork-like stick.

STAYS
Rods of hazel or willow used with scollops to secure thatch.

STOOK
A group of sheaves of grain stood on end in a field to dry.

STRAW
The stalks of threshed grain, especially of wheat, rye, oat or barley.

SÚGÁN
Straw or hay rope used to tie sod foundation and base layers of thatch to roof structure.

THATCHING FORK
Metal or wooden implement, having two prongs at one end, used to thrust bundles of straw into old thatch.

THRESHING
A method of removing grain from straw. Threshing can be carried out manually or mechanically.

VERNACULAR
Native or indigenous form of building generally using locally available materials.

WATTLE
Interlaced rods of hazel and willow used to form partition walls, and the support structure for the thatch base layer.

READING LIST & INFORMATION ON GRANTS

Danaher, Kevin: *Ireland's Vernacular Architecture*, Mercier Press, Cork, 1975.

Gallagher, Joseph and Stevenson, Greg: *Traditional Cottages of Donegal*, Under the Thatch Ltd, UK, 2012.

Pfeiffer, Walter and Shaffrey, Maura: *Irish Cottages*, Artus Books, London, 1990.

Reeners, Roberta: *A Wexford Farmstead: the Conservation of an 18th-century Farmstead at Mayglass*, The Heritage Council, Ireland, 2003.

Shaffrey, Patrick and Maura: *Irish Countryside Buildings: Everyday Architecture in the Rural Landscape*, The O'Brien Press, Dublin, 1975.

Wyse Jackson, Peter: *Ireland's Generous Nature: The Past and Present Uses of Wild Plants in Ireland*, Missouri Botanic Garden Press, St. Louis, 2014.

SURVEYS

The National Inventory of Architectural Heritage Building Survey, www.buildingsofireland.ie

Built Heritage, Northern Ireland Environment Agency.

GRANTS FOR THATCHED BUILDINGS IN IRELAND

Grants are awarded on the basis of criteria set by the granting body.

Department of the Environment, Heritage & Local Government Housing Grants,
Government Buildings, Ballina, Co. Mayo.
Tel: 096 24200
LOCAL 1890-30-50-30
www.environ.ie

The Heritage Council, The Bishop's Palace, Church Lane, Kilkenny
Tel: 056 777 0777
mail@heritagecouncil.com
www.heritagecouncil.ie

Local Authority Grants: contact each local authority for further information.

ACKNOWLEDGEMENTS

Many thanks to all those who helped me in the Department of Arts, Heritage and the Gaeltacht, in particular Jane Wales who furnished me with a list of listed thatched properties in Ireland, also thanks to Jacqui Donnelly, the Architectural Conservation Advisor. A debt of thanks is also due to Eoin Fegan and his wife Helen Carr who translated map co-ordinates into 'sat nav' speak, without whom I would have literally been lost. Thanks to the men and women whom I met on the roads of this country who love and keep these properties in beautiful condition, often in straightened financial circumstances. Thanks to Jonathan for sharing the journey. Finally a debt of thanks to the wonderful group of people at The O'Brien Press, not least my patient editor Susan Houlden.

PHOTOGRAPH CREDITS

The majority of the images used in this book are the author's own. Others are reproduced with kind permission of: pp164-5 images courtesy of Emily Mannion.The author received permission from the OPW to photograph the interior of the Dwyer McAllister Cottage, Wicklow (pp66-67); the Swiss Cottage, Cahir, County Tipperary (pp94-5); and Patrick Pearse's Cottage, Galway (pp140-1). The interior of Swiss Cottage, p95 bottom, © National Monuments Service, Department of Arts, Heritage and the Gaeltacht. Glencolmcille Folk Village in the snow, p150 top, Margaret Rose Cunningham.

All images photographed by the author have been shot on digital SLR cameras using numerous lenses, manual focus and automatic. I have used 2.8/70-200mm, 2.8/18-105 mm for most of the images.

First published 2015 by
The O'Brien Press Ltd,
12 Terenure Road East, Rathgar,
Dublin 6, Ireland.
Tel: +353 1 4923333; Fax: +353 1 4922777
E-mail: books@obrien.ie.
Website: www.obrien.ie

ISBN: 978-1-84717-692-9

8 7 6 5 4 3 2 1
19 18 17 16 15

Printed and bound in Poland by Białostockie Zakłady Graficzne S.A.
The paper in this book is produced using pulp from managed forests.